SPIDERS Play Tricks!

BY MEGHAN GREEN

ANIMAL TRICKSTERS

Gareth Stevens PUBLISHING

Please visit our website, www.garethstevens.com. For a free color catalog of all our high-quality books, call toll free 1-800-542-2595 or fax 1-877-542-2596.

Cataloging-in-Publication Data

Names: Green, Meghan.
Title: Spiders play tricks! / Meghan Green.
Description: New York : Gareth Stevens Publishing, 2025. | Series: Animal tricksters | Includes glossary and index.
Identifiers: ISBN 9781538293607 (pbk.) | ISBN 9781538293614 (library bound) | ISBN 9781538293621 (ebook)
Subjects: LCSH: Spiders–Juvenile literature. | Spiders–Behavior–Juvenile literature.
Classification: LCC QL458.4 G782 2025 | DDC 595.4'4–dc23

Published in 2025 by
Gareth Stevens Publishing
2544 Clinton Street
Buffalo, NY 14224

Designer: Andrea Davison-Bartolotta
Editor: Jennifer Lombardo

Photo credits: Cover, pp. 1 Manjunath acharya/Shutterstock.com; series art (background texture) Kriangkrai Namtongbai/Shutterstock.com; series art (web texture) ALEX S/Shutterstock.com; p. 4 common human/Shutterstock.com; p. 5 (main) Pogorelova Olga/Shutterstock.com; p. 5 (spider) BlueRingMedia/Shutterstock.com; p. 5 (sun) Studio Barcelona/Shutterstock.com; p. 7 sivananthan2001/Shutterstock.com; p. 9 (main) blickwinkel/Alamy Stock Photo; p. 9 (inset) Gchapel/Shutterstock.com; p. 11 David Havel/Shutterstock.com; pp. 12, 13 Ondrej Michalek/Shutterstock.com; pp. 15, 17 Pong Wira/Shutterstock.com; p. 16 Lukas Gojda/Shutterstock.com; p. 18 Zety Akhzar/Shutterstock.com; p. 19 Lee Hua Ming/Shutterstock.com; p. 21 martinusnovedwin/Shutterstock.com.

Printed in the United States of America

CPSIA compliance information: Batch #CSGS25: For further information contact Gareth Stevens at 1-800-542-2595.

CONTENTS

Words in the glossary appear in **bold** type the first time they are used in the text.

HELPFUL SPIDERS

Many people around the world are scared of spiders. However, most spiders are actually helpful to people, not harmful! Spiders eat the bugs that eat our food. Without spiders, there might not be enough food for everyone to eat.

There are more than 45,000 known species, or kinds, of spiders in the world. Many have cool tricks that help them survive. These tricks can help them hide from predators, catch **prey**, or keep themselves safe in other ways.

Spiders are an important part of the **ecosystem**.

THE TWIG MIMIC SPIDER

There are 43 species of spiders that can make themselves look like twigs, or small branches. This **camouflage** helps them hide from predators. The main part of the spider's body is called its abdomen. The twig mimic spider's abdomen is long and pointed.

When it needs to hide, the spider folds its legs up along the sides of its abdomen. It stands very still on a tree branch. To a predator, it looks just like a twig sticking out of the bigger branch!

This spider mimics, or looks like, a twig.

FUN FACT

Some people also call these spiders "tree stump mimics" even though they're much smaller than tree stumps.

THE DIVING BELL SPIDER

Most spiders live on land. However, the diving bell spider has a trick that lets it live underwater almost all the time. First, it spins a web that looks like a sheet. Then it traps a bubble of air, dives down with it, and releases it under the web.

The spider lives inside this air bubble for most of the rest of its life. If it runs out of air, it goes to the surface to get more. When the spider needs to move around outside the bubble, it can put another air bubble around its abdomen, which is where its lungs are.

The diving bell spider is named after a machine humans use to let them dive safely.

DIVING BELL

THE GOLDEN WHEEL SPIDER

The golden wheel spider lives in the Moroccan desert, where there are large sand hills called dunes. The spider makes its home inside these dunes. However, it isn't always safe inside its home. Spider wasps, the spider's main predator, can move up to 2.6 gallons (10 L) of sand to get at the spider.

To avoid the wasp, the golden wheel spider sometimes needs to make a quick getaway. It does this by curling its legs up around its body and cartwheeling down the sand dunes.

The golden wheel spider tucks in its long legs to roll its body over and over down a sand dune.

FUN FACT

The golden wheel spider is most **vulnerable** to wasps when it's in the middle of building its home. Wasps can't dig deep enough to get to it when it's finished.

THE NET-CASTING SPIDER

Most spiders spin a web and wait for prey to fall into it. However, the net-casting spider takes a more **proactive** approach. After night falls, it comes out and knits a web. It hangs upside down, holding the web between its front legs.

When an insect comes along, the net-casting spider throws the web over it, like a fisherman throwing a net into the ocean. Then the spider drops down and grabs the prey it has trapped.

The net-casting spider is also called the ogre-faced spider.

FUN FACT

Net-casting spiders have the biggest eyes of any spider on Earth. These help them see and grab their prey in the dark.

THE TRAPDOOR SPIDER

All spiders can make silk, but not all of them spin webs the same way. The trapdoor spider uses its silk to make an underground tunnel web. This tunnel has more than one way out. If the spider is in danger, it can quickly leave out of any of its exits.

However, the front door is the one that's special. From outside, the door is hidden, so prey can't easily see it. On the back, the door is covered in webbing. The trapdoor spider can feel **vibrations** through this webbing.

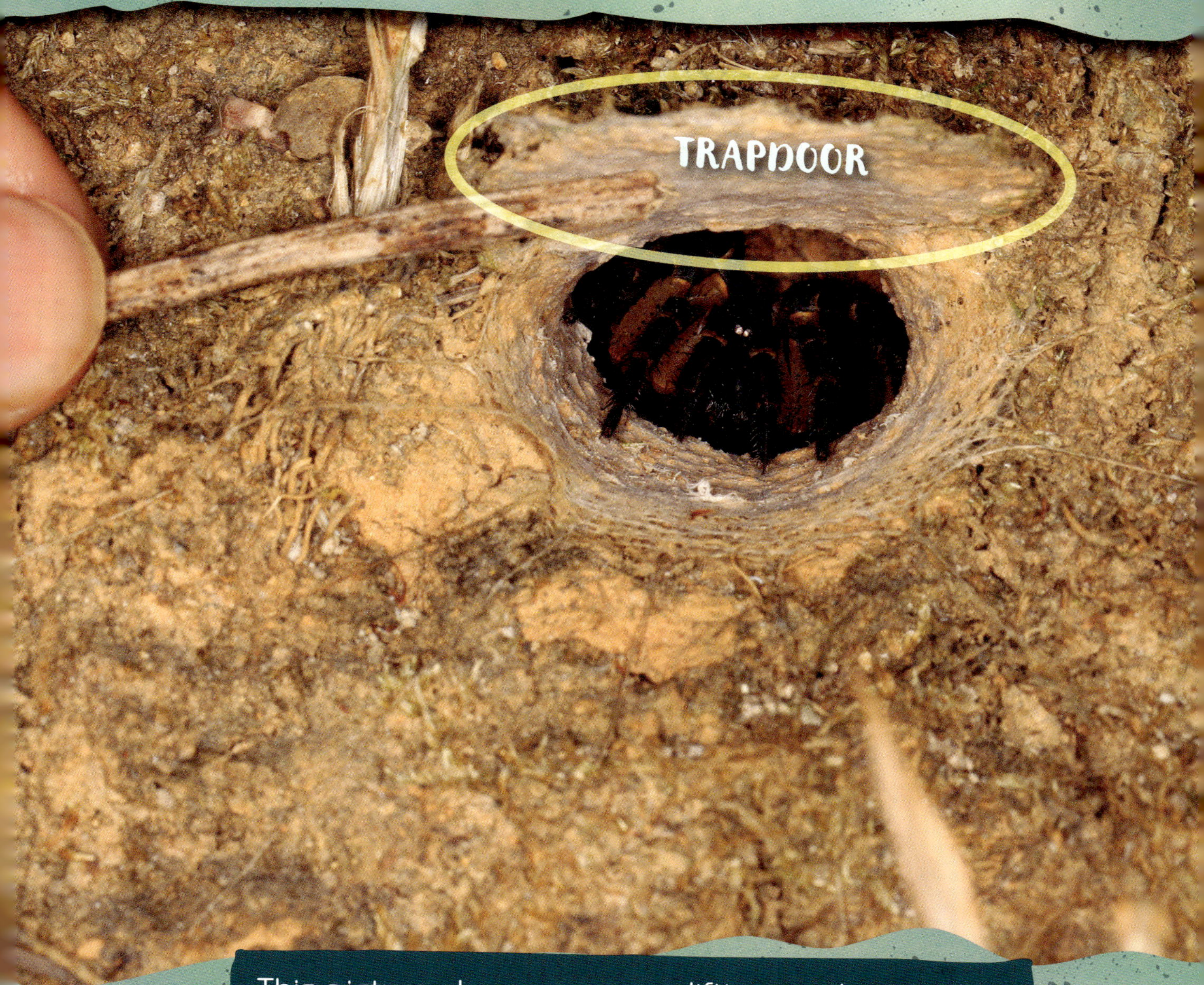

This picture shows a person lifting up the trapdoor to show the spider in its underground home.

CAUGHT!

The trapdoor spider waits without moving behind its trapdoor. When prey steps on the trapdoor, the web vibrates. When the spider feels those vibrations, it jumps! It bursts out through the trapdoor and grabs its prey, then drags it back into the tunnel to eat it.

Trapdoor spiders mostly eat insects such as grasshoppers and crickets. However, they're known to also eat frogs, mice, and even baby birds. They like to live where it's warm. In the United States, they can be found south of Virginia.

This picture shows what a trapdoor spider's home looks like from the outside. This spider is the only one in the world that makes a home like this.

FUN FACT

The trapdoor spider can keep itself safe from predators by holding the trapdoor closed when an animal is trying to get in.

THE ANT MIMIC CRAB SPIDER

Plants aren't the only things some spiders can mimic. The ant mimic crab spider preys on weaver ants. The spider also looks a little like this kind of ant. It has the same orange color, and its body is shaped a little like the ant's.

The spider has an extra trick up its sleeve to get its prey to come close enough to catch. Ants have six legs and two antennae. Spiders have eight legs and no antennae. To fool the ants, the spider raises its front two legs above its head to look like antennae!

This picture shows an ant mimic crab spider (left) catching a weaver ant (right).

THE LADYBIRD MIMIC SPIDER

Another spider that can mimic an animal is the ladybird mimic spider. However, this isn't because the spider preys on ladybirds. It's because ladybirds taste bad to birds! Spiders, on the other hand, taste delicious to them.

When a bird sees a ladybird mimic spider, the bird will often not even try to eat the spider. It has learned from eating ladybirds in the past that the colors and spots mean the meal won't taste good. With this trick, ladybird mimic spiders can hide in plain sight.

FUN FACT

In the United States, ladybirds are more commonly known as ladybugs.

Most spiders have eight eyes, but the ladybird mimic spider has only four.

GLOSSARY

camouflage: Colors or shapes on animals that allow them to blend in with their surroundings.

ecosystem: A natural community of life that includes living and nonliving features.

proactive: Taking action instead of waiting.

prey: An animal that is hunted by other animals for food.

vibration: A rapid movement back and forth.

vulnerable: In an unsafe position.

FOR MORE INFORMATION

Books

London, Martha. *Spiders*. Lake Elmo, MN: Focus Readers, 2021.

Lundgren, Julie K. *Spiders*. New York, NY: Crabtree Publishing Company, 2022.

Nicholas, Christopher. *Spiders*. London, UK: Sequoia Kids Media, 2023.

Websites

DK Find Out!: Spiders
www.dkfindout.com/uk/animals-and-nature/arachnids/spiders
Learn more facts about some of the coolest spiders out there.

Kiddle: Spider Facts for Kids
kids.kiddle.co/Spider
Read more about spiders and take a look at some up-close photos—if you dare!

YouTube: Reinventing the Wheel: 5 Species that Roll
www.youtube.com/watch?v=KIE7iET7yBo
Check out five kinds of animals that roll out of danger, including the golden wheel spider.

Publisher's note to educators and parents: Our editors have carefully reviewed these websites to ensure that they are suitable for students. Many websites change frequently, however, and we cannot guarantee that a site's future contents will continue to meet our high standards of quality and educational value. Be advised that students should be closely supervised whenever they access the internet.

INDEX